Get Coding with Python™

By Jenna Vale

Gareth Stevens
PUBLISHING

Please visit our website, www.garethstevens.com. For a free color catalog of all our high-quality books, call toll free 1-800-542-2595 or fax 1-877-542-2596.

Library of Congress Cataloging-in-Publication Data
Names: Vale, Jenna, author.
Title: Get coding with Python / Jenna Vale.
Description: Buffalo, New York : Gareth Stevens Publishing, [2024] | Series: Get coding! | Includes index. | Audience: Grades 4-6
Identifiers: LCCN 2023011460 (print) | LCCN 2023011461 (ebook) | ISBN 9781538288641 (library binding) | ISBN 9781538288634 (paperback) | ISBN 9781538288658 (ebook)
Subjects: LCSH: Python (Computer program language)–Juvenile literature.
Classification: LCC QA76.73.P98 V355 2024 (print) | LCC QA76.73.P98 (ebook) | DDC 005.13/3–dc23/eng/20230317
LC record available at https://lccn.loc.gov/2023011460
LC ebook record available at https://lccn.loc.gov/2023011461

Published in 2024 by
Gareth Stevens Publishing
2544 Clinton Street
Buffalo, NY 14224

Portions of this work were originally authored by Patricia Harris and published as *Understanding Coding with Python*. All new material in this edition authored by Jenna Vale.

Designer: Leslie Taylor
Editor: Megan Kellerman

Photo credits: Cover Trismegist san/Shuttertstock.com; cover & series art (coding background) BEST-BACKGROUNDS/Shutterstock.com; p. 5 Doc Searls/https://commons.wikimedia.org/wiki/File:Guido_van_Rossum_OSCON_2006.jpg; p. 7 Robvanroo/https://commons.wikimedia.org/wiki/File:Entree_CWI.jpg; p. 9 Prostock-studio/Shutterstock.com; p. 11 Monkey Business Images/Shutterstock.com; p. 13 dreamerve/Shutterstock.com; p. 15 guteksk7/Shutterstock.com; p. 17 Szasz-Fabian Ilka Erika/Shutterstock.com; pp. 22, 25 sundatoon/Shutterstock.com; p. 27 IB Photography/Shutterstock.com; p. 29 wavebreakmedia/Shutterstock.com.

Printed in the United States of America

CPSIA compliance information: Batch #CS24GS: For further information contact Gareth Stevens at 1-800-542-2595.

CONTENTS

Words in the glossary appear in **bold** the first time they are used in the text.

Learning the Language

Many devices we use daily have computers in them: cars, refrigerators, and even toys. The computers in these devices can be very different from each other. They can be different in size, components, and even in what language they're programmed in.

With so many **programming languages** out there, how do you choose which one to learn? One way is to choose a language that can be used for writing a wide variety of programs, such as Python. A Dutch programmer named Guido van Rossum created the Python programming language. He wanted Python to be easy to use and understand so more people could learn to code.

Guido van Rossum, seen here, began **developing** Python in the late 1980s.

Monty Python

Van Rossum named Python after the 1970s British comedy show *Monty Python's Flying Circus*. Python coders are even encouraged to use quotes, or lines, from the show when coding in Python! Van Rossum said he chose Python because he needed a name that was "short, unique, and slightly mysterious." He also read *Monty Python* screenplays while developing the language.

Easier Than ABC

Before Python, Guido van Rossum helped develop a programming language called ABC. It was meant to be understandable to people who weren't computer scientists. However, Van Rossum felt ABC was still too hard to use. He developed Python to have the best features of ABC without having the same problems.

Python is a standard text-based, or word-based, programming language. That means it uses words and special characters to create code. It can have code grouped into **loops** or in other ways. It has some special **modules** built in to help save time and make coding less repetitive too.

This is the Centrum Wiskunde & Informatica, a national research institute for mathematics and computer science in the Netherlands, where Van Rossum worked with a team to create the ABC programming language.

Lowercase Keywords

Code entered into the computer is called input, and the results displayed on the computer screen are called output. With ABC and Python, inputting certain **keywords** always provides the same output. ABC had keywords in uppercase, which was confusing even to Van Rossum! He developed Python to always take keywords in lowercase instead.

Basic Coding Rules

Computer programming is all about following rules no matter what language you're using. You might think that sounds a lot like playing a game, and it can be just as much fun.

Rule 1: Coders must know what they want the computer to do and write a plan.

Rule 2: Coders must use special words to have the computer take input, make choices, and take action.

Rule 3: Coders need to think about what tasks can be put into a group.

Rule 4: Coders need to use **logic** with AND, OR, NOT, and other key logic words.

Rule 5: Coders must explore the **environment** and understand how it works.

Writing out a plan will help you understand how you need to break down each instruction for the computer.

Free Play

You can start learning the rules of programming with Python at www.python.org. Python is free to **download**, and you can start playing with it right away! Guido van Rossum may have started Python, but many people have added to it over the years. Python is open source, which means anyone can make and share changes to the language.

Print on Demand

One of the basic **statements** you can input is the "print()" **function**. Type what you want to display on the computer screen in between the parentheses, then hit "Enter" or "Return" on your keyboard. This will show whatever you've told Python to print on the computer screen.

If you want Python to display the number 5, type the code "print(5)." Then the output "5" will appear on your screen—it's that easy! Math operations are built into the Python language. So, to find out the answer to 31 times 56, type the code "print(31*56)." Python will then display "1736" on the screen.

A Lot of Character

Anything you type into Python is called a character. Characters include letters, quotation marks, numbers, and even spaces. However, pay attention to how certain characters, such as parentheses and quotation marks, are used. These characters work differently in Python than they do in regular writing.

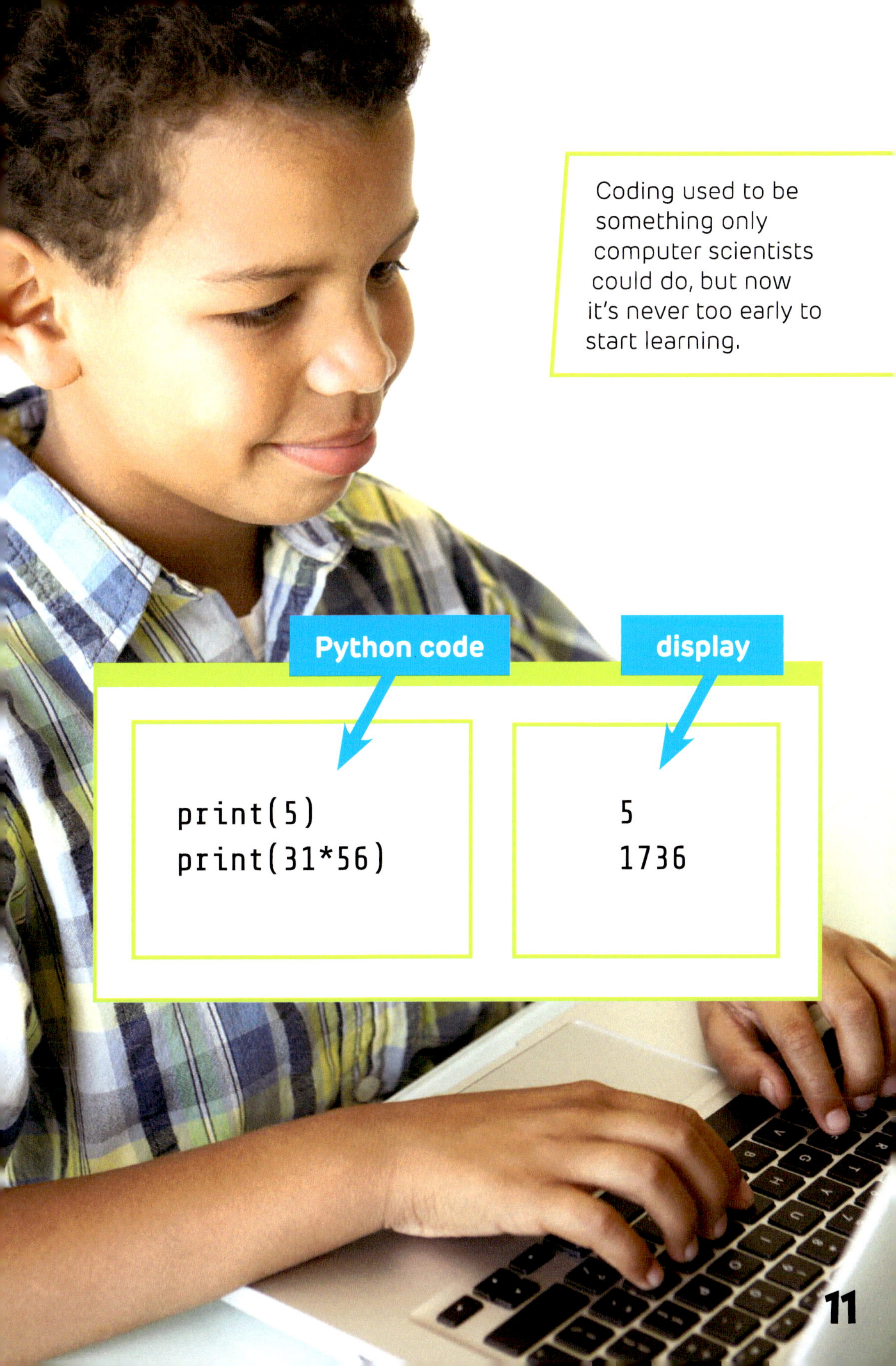

Coding used to be something only computer scientists could do, but now it's never too early to start learning.

Python code

```
print(5)
print(31*56)
```

display

```
5
1736
```

Displaying a word or sentence is the same "print()" function with one extra step. Inside the parentheses, you need to add quotation marks around the text you want to print. You have two options for writing this kind of code:

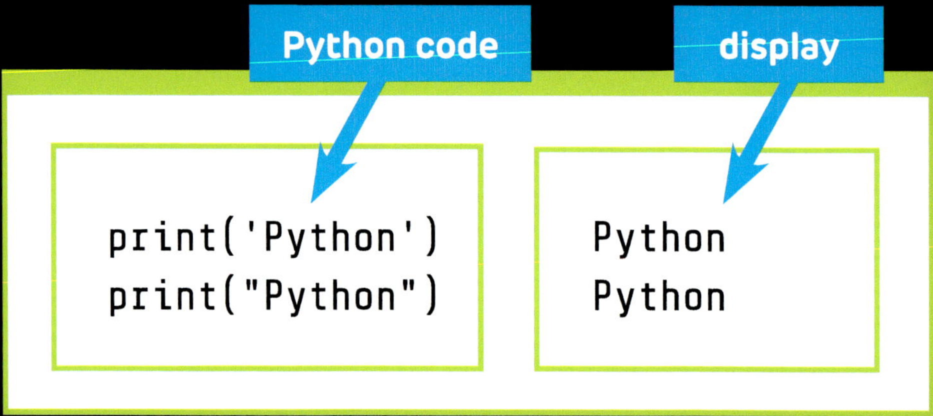

Both of these statements will display the word "Python" (without quotation marks) on your screen. You can use single quotation marks or double quotation marks on either side of your text. Either of these options will work, but remember to be consistent in your code. Pick one option, and use it for all your print functions.

Syntax Error

If you make any mistakes while coding, Python will let you know. For example, typing "print 'Python'" will create a "SyntaxError" that reads "Missing parentheses in call to 'print'. Did you mean print(...)?" Syntax is the order in which you put characters and words. Python is great for checking common mistakes as you code.

It's important to be consistent and keep things simple. Which quotation marks would work best for you?

IDLE Shell

Coders write programs using a text editor. Most computers come with a text editor, but there are many you can download. The editor allows you to write code, but it doesn't show the results.

If you decide to download Python from www.python.org, an editor called IDLE Shell will download along with the language. This is the standard development environment for Python. You'll be able to save your code by selecting "File" from the top left corner and then "Save As." Python file names end in .py. You can also use free online text editors to do basic coding exercises.

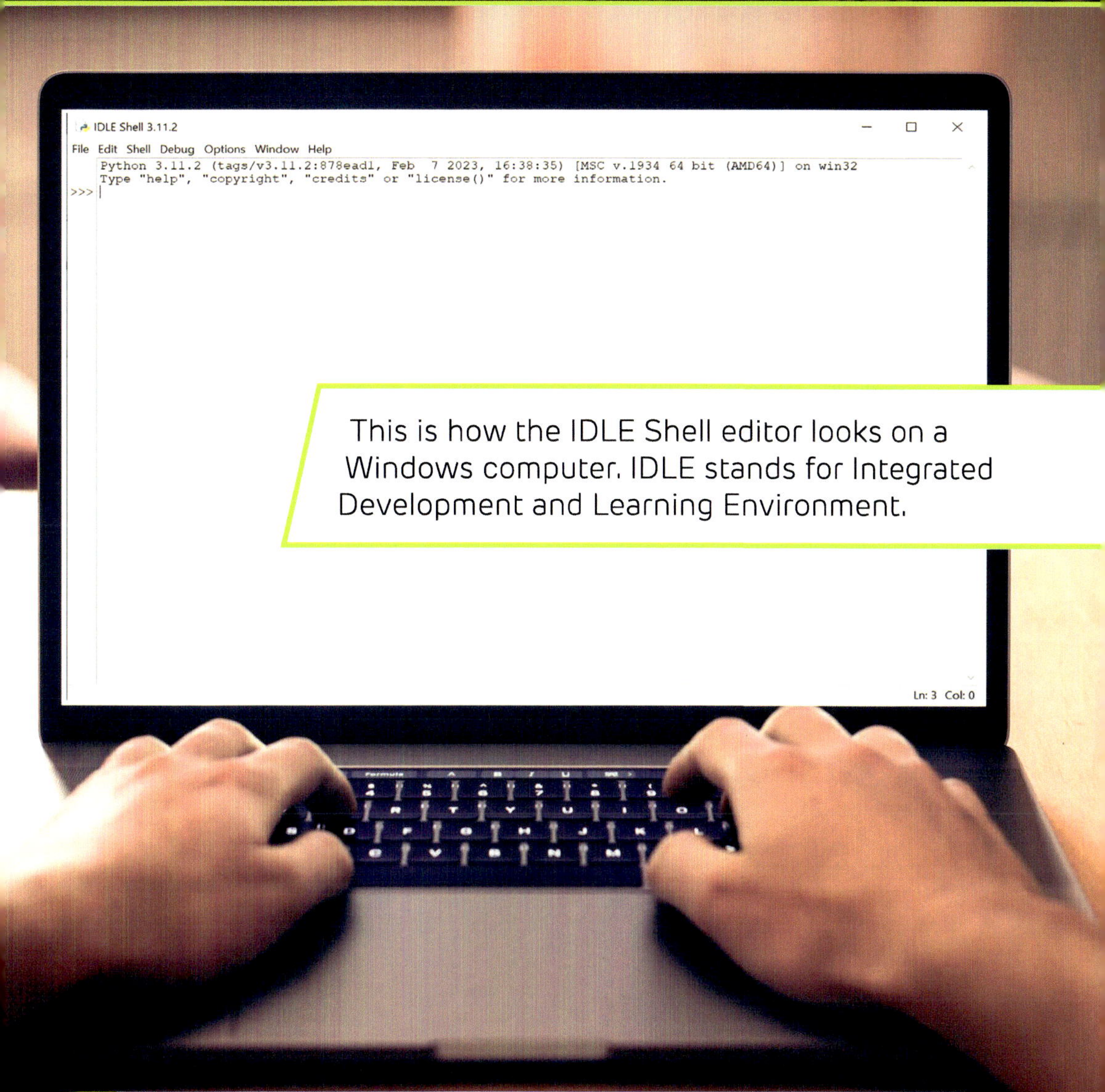

This is how the IDLE Shell editor looks on a Windows computer. IDLE stands for Integrated Development and Learning Environment.

Built-in Math

Remember how we used the print function to display the answer to a math problem earlier? You can also use the IDLE Shell as a calculator! Input any math problem to have Python calculate the answer for you. That means if you type 4+4 and hit "Enter" or "Return," 8 will show up on the next line.

Your First Program

The Hello World program is the first program many beginner coders learn. The plan is to tell the computer to print to the screen the words "Hello, World!" In Python, this simple program looks like this:

```
#A simple program to print out Hello, World!
print('Hello, World!')
```

Notice the first line starts with #, which is called a hash. That is the character for a comment in Python. Comments are for making notes in your program without affecting your code. If your comment runs for more than one line, make sure to add # at the beginning of each line.

```
#A simple program to print out Hello, World!
print('Hello, World!')
```

Python code

```
Hello, World!
```

display

Writing a Hello World program can be very exciting! It's the first of many lessons you'll learn while coding.

Strings

Any text you print in Python is called a string. This term is used in many programming languages. When you add quotation marks around text in Python, this lets Python know the text is a **sequence** of words meant to be read. This means Python won't misunderstand your text as a number, command, or variable.

What Is Your Name?

Another simple program you can write makes it look like Python is talking to the user. This is a useful program because developers often need users to interact with programs to collect data or produce certain outcomes. In the code, you'll ask the user to enter their name and then have Python print the user a message with the name included.

This program requires the input function, input(). When you use this function, the program will stop and wait for the user to provide input. When the user does, Python will receive whatever they type as a string. The program will then continue running.

The first table shows what the code will look like. The second table shows what the program will look like when it's running.

GetName.py

```
#A program to ask a user their name and say hello
name = input('What is your name?\n')
print('Hello', name)
```

1.) The top line is a comment that describes the program. It begins with a hash character and is only one line.

2.) Code to define the input function. Adding "\n" to the end of the question means Python will produce a new line for the user to write their input on after printing the question.

3.) The program will continue with this print function. The name input was defined earlier, so it's included in the code here.

Running the Program

Output	
`What is your name?`	How the question will appear to the user.
`Jessie`	The new line where the user will type their name.
`Hello Jessie`	How Python's greeting will appear to the user.

Dialog Boxes

When input from a user is needed, a program will often use a dialog box for the input. A dialog box is a separate window where the user can type things in, click on buttons to agree or disagree on something, or receive new directions from the program.

A Longer String

Let's build off this program. Coders can use a program already written and add new code. This makes coding quicker. Usually, they save the new program under a new file name so they have both the original program and the new one. If you're using the IDLE Shell editor, this means you'll choose "Save As" under "File" to make sure you keep both the original program and the new one.

We'll call this program "NameAge.py." The code will show input for a name and an age. Have a look at the code on the next page to see what you need to do.

Operators

When creating longer strings, you'll sometimes use even more types of characters than we've used so far. One kind of character is the operator. An operator is a character such as +, *, or /. These can be used in math problems, but they can also be used with strings.

Give this code a try to practice using the input function. Create more input variables and have Python ask even more questions!

NameAge.py

```
1  name = input('What is your name?\n')
2  print('Hello', name)
3  age = input('How old are you?\n')
4  print('Hi again', name+'!', 'I see you are', age,
   'years old.')
```

1.) This is the same code we used in the last program.

2.) This is the same code we used in the last program.

3.) The new "age" input is written the same way the "name" input was.

4.) The new string we want Python to print uses a few different parts. The "+" operator after the name input will let Python add an exclamation point after the name. Pay attention to how the colors and commas work to show you the different parts of the string.

Running the Program

Output	Explanation
`What is your name?`	Python will ask the user their name as before.
`Jessie`	Jessie will give their name.
`Hello Jessie` `How old are you?`	Python will greet Jessie, then ask Jessie's age on a new line.
`10`	Jessie will give their age.

`Hi again Jessie!`
`I see you are 10`
`years old.`
	Python will say hi again and note Jessie's age.

Learn with Turtle

Even though Python is text-based, you can use it to create graphic, or picture, results. One way to learn this is through the Turtle graphics module. You write code and watch the turtle draw the coded actions in a special window. Think of the turtle as a robot you control. Turtle helps you begin to think about the special sentence structure and format needed in Python. It's also fun!

You have to call, or open, the special module at the start of your program. This program requires the use of loops and variables. "Length" is an example of a variable because it can be replaced by different numbers.

This shows the code for having the turtle draw a square on the screen. In the code, you must first call the graphics screen, define the turtle variable (t), and then draw your program. But this code is a little long, right?

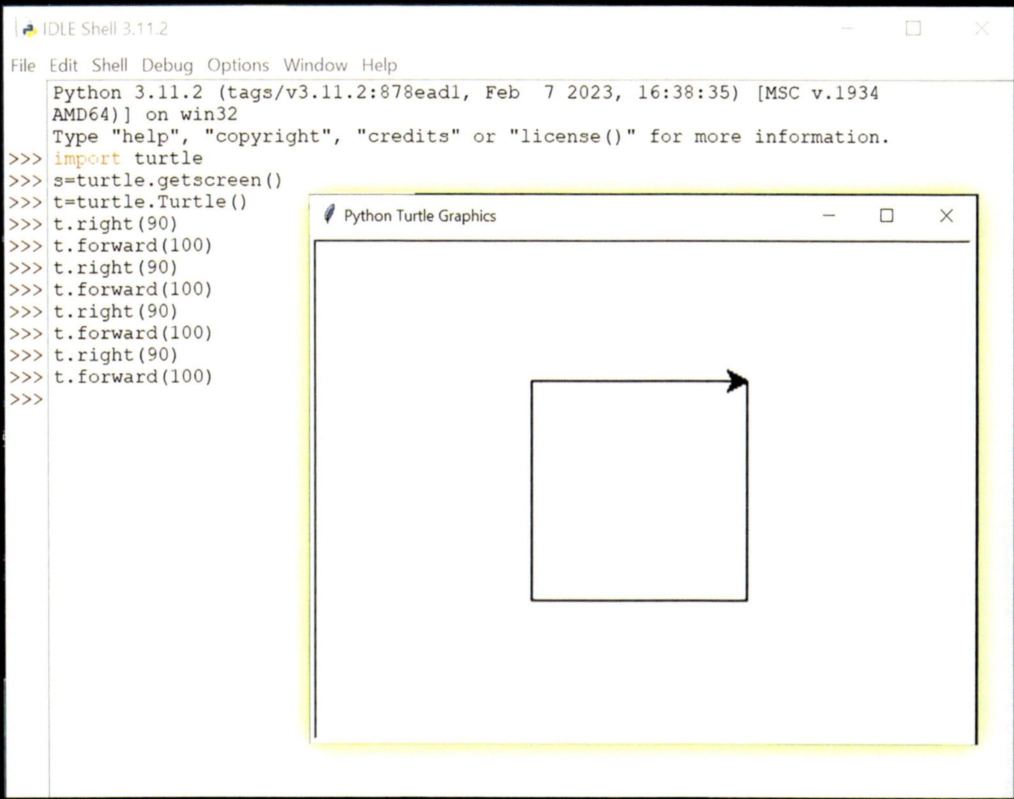

Come Out of Your Shell!

You can also make fun changes to the output for your turtle program. You can change the way the turtle looks, add a different background color, and change the length and width of the line the turtle draws. Get creative—that's what turtle graphics are for!

The code to make the turtle draw a square on the last page listed out every single step. Now that we've been coding for a while, we know we can make the code shorter by using a loop. Because a square is four sides and four angles, we can group the turtle's entire journey into one loop.

You'll use a "for" loop to do this. In Python, a "for" loop is a loop of code that you want to repeat for a set number of times. In our case, we will repeat the turns and forward movement four times. See the code on the next page to see what this looks like.

The While Loop

Another kind of loop in Python is the "while" loop. Unlike a "for" loop, a "while" loop will not repeat for a set number of times. Instead, this loop of code will run while a certain event is happening and will end once that event ends.

This code is much easier to read. The loop begins on the fourth line with "for." Notice that the third line of code also changed the way the turtle looked—in this program, it looks like a real turtle!

Make a Statement

All the coding examples we've looked at so far have been made up of simple statements. A simple statement exists on one logical line. One kind of simple statement we haven't used yet is the break statement. This statement can be used only inside a "for" loop or "while" loop. A break statement will let the program exit out of the loop before it ends.

You can also use compound statements in Python. Compound statements are statements that have groups of statements contained within them. The more kinds of statements you learn about in Python, the more you'll be able to do!

Python has grown a lot over the years, but it's still a simple language to learn and use.

Reading Python

There are many kinds of statements to learn with Python. Don't worry if learning about them seems hard at first! Unlike some other programming languages, Python is very similar to regular written English. There are plenty of lessons, guides, and modules to help you code with Python.

Before You Know It

Python is one of the most popular programming languages being used today. It's a general-purpose language, which means it can be used to write many different types of programs. Whether you want to learn how to build a website, organize data, or create programs, Python is an excellent language to code in.

You can also create graphics with Python. Adding graphics to your code can make your programs look nicer and show off your creativity. They also make programs more fun to use. Whatever you'd like to code, learn the basics of programming with Python—you'll be coding before you know it!

Remember the Basics

Don't forget to keep practicing the basics as you learn Python. Write out a simple plan before you begin coding. Think about how parts of your code can be grouped. Add comments to help you remember what different parts of your program are supposed to do. Building on the basics will help your coding in the long run.

Python is a great language to learn coding with and can help you move on to more advanced languages in the future!

GLOSSARY

develop: To create something and work to improve it over time.

download: To move or copy a file or program from one computer or device to another.

environment: The combination of computer hardware and software that allows a user to perform tasks.

function: In Python, a block of code that will run only when instructed to do so.

keyword: A word that has a special meaning and is used for a specific purpose.

logic: A proper or reasonable way of thinking about or understanding something.

loop: A process that starts over once it reaches the end.

module: A Python file with a premade set of functions that can be saved and used to save time while coding.

programming language: A coding language designed to give instructions to a computer.

sequence: A group of things in a certain order.

statement: In Python, an instruction that can be understood and carried out by the computer.

FOR MORE INFORMATION

Books

Briggs, Jason R. *Python for Kids, 2nd Edition: A Playful Introduction to Programming*. New York, NY: No Starch Press, 2022.

McCue, Camille. *Getting Started with Coding: Get Creative with Code!* Hoboken, NJ: Wiley, 2020.

Foster, Patricia. *Python Programming for Beginners: A Kid's Guide to Coding Fundamentals*. Emeryville, CA: Rockridge Press, 2020.

Websites

Code.org
www.code.org
An educational nonprofit focused on providing computer science tools for grades K–12.

Python
www.python.com
The official website for Python. Download Python for any device and learn more about how to code.

Real Python
www.realpython.com
An online learning platform featuring lessons, tips, quizzes, and more to help you practice coding in Python.

Publisher's note to educators and parents: *Our editors have carefully reviewed these websites to ensure that they are suitable for students. Many websites change frequently, however, and we cannot guarantee that a site's future contents will continue to meet our high standards of quality and educational value. Be advised that students should be closely supervised whenever they access the internet.*

INDEX